CATS

by Nina Leen

Holt, Rinehart and Winston / New York

Copyright © 1980 by Nina Leen
All rights reserved, including the right to reproduce
this book or portions thereof in any form.
Published simultaneously in Canada by Holt, Rinehart
and Winston of Canada, Limited.
Printed in the United States of America
10 9 8 7 6 5 4 3 2 1

Library of Congress Cataloging in Publication Data
　Leen, Nina　　Cats.

Summary: Text and photographs describe the
characteristics and behavior of cats and introduce
domestic and wild members of the cat family.
1. Cats—Behavior—Juvenile literature.　2. Felidae
—Juvenile literature. [1. Cats. 2. Felidae]
I. Title. SF446.5.L43　599′.74428
79-22137　ISBN 0-03-052331-1

ACKNOWLEDGMENTS

My thanks to everybody who let me observe and take pictures of their cat's behavior. I also appreciate the help given me by Time-Life Photo Lab. Lack of space makes it impossible to list all books and scientific material made available to me. The picture of the cat nursing a puppy on page 66 was taken by Mrs. Martha Raynor.

FOREWORD

Despite scientific observation and countless studies, the best experts of cat behavior are people who live with cats, learn to understand their language and their various likes and dislikes, and respect them.

Cats have unique personalities: they are nobody's obedient subject; they don't try to flatter the human ego; they don't have masters, they have friends; and they expect those friends to treat them with due affection. When a cat enters our lives, it promptly enjoys undisputed privileges: wherever the cat selects to rest nobody will dare to chase it away—even if it is a favorite chair, needlepoint pillow, or heirloom bedspread. It can come and go as it pleases, move around the house without restrictions, and play with all kinds of objects. In return, the cat shows appreciation by its warm, dignified tenderness and by patiently forgiving many human shortcomings. People who dislike cats often insist that cats are incapable of love and devotion—a mistake they make only because they don't know cats.

Somebody once said that a home is not a home without a cat, and millions of people will agree.

N. L.

Wild Relatives

Felidae—the family of cats—includes large and small wild cats, as well as the domestic cats, *Felis catus.* All cats have common characteristics in anatomy and behavior. Cat language is universal—wild or domestic, they purr, signal with their tails, move their ears, have a large vocabulary of sounds, and can convey their emotions with their flexible bodies. Because of illegal slaughter for their fur and the continued destruction of their habitats, these beautiful wild animals may soon disappear. All the wild cats in this book are endangered species.

The baby snow leopard will grow up to be a big, furry, beautiful cat. It lives in the sparsely populated mountain regions of Asia.

The black-footed cat of southern Africa, the smallest of the wild cats, occasionally mates with domestic cats. Shy and nocturnal, this cat is rarely seen.

When cornered, it puts up a tough front.

This young puma is a member of a large cat family. Pumas have many names: catamount, mountain lion, cougar, panther are some. They inhabit a wide range from Canada to South America. Young pumas are spotted. As they grow older, the spots fade away.

The margay may appear in the southern part of the United States, but its habitat is in Mexico and Central and South America. It is a close relative of the ocelot, but smaller.

Cheetahs live in parts of Africa and Asia. In some ways they are different from other cats: they hunt during the day, run short distances faster than any other four-legged animal in the world, partly retract their claws, and are easily tamed.

The clouded leopard of Asia spends the day in jungle trees. At night it comes down to hunt.

Worship and Superstition

An Egyptian Bronze Cat:
Ancient Egypt was the home of the first domesticated cats. The Egyptians admired, bred, and worshiped them. Laws protecting cats were severe: a person guilty of killing a cat had to pay with his life. When a cat died it was embalmed, mummified, and given an elaborate burial.

The Black Cat:
Of the many superstitions about cats, most are about black cats. In the middle ages they were thought to be in league with the devil and witches. Today, beliefs differ—in some countries a black cat means trouble; in others, good luck, happiness, and riches.

It is not true that cats' eyes emit light rays, thus making it possible for them to see in total darkness. What makes the eyes "shine" at night is a mirrorlike layer of cells behind the retina that reflects any existing light.

Senses

SIGHT

Cats have excellent eyesight. On moonless nights they can move around, stalk prey, and see objects and movements invisible to people and most animals. Their sight is sensitive in the dimmest light—but not in absolute darkness.

In a bright light the pupils of the eyes become narrow vertical slits.

Cat eyes are exceptionally large. The cat watches for a long time without blinking.

When the light is dim, the pupils open wide to let the light pass to the retina.

TASTE

The cat has definite preferences in its eating habits. If the food is not to its taste, it will refuse to eat. It does not gulp down food but savors it. When hunger is satisfied, food is left for a later snack.

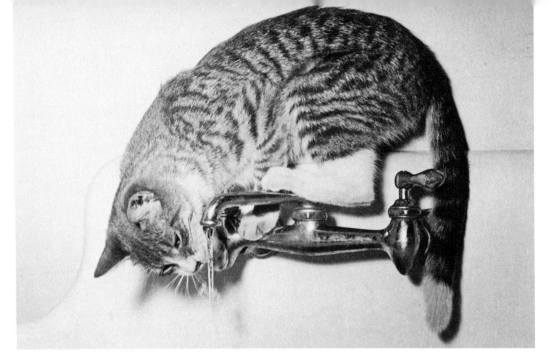

When the water bowl is empty, the cat takes a drink from the faucet.

It is commonly believed
that all cats like milk.
Adult cats think differently—
some may like it,
others never touch it.

TOUCH AND AFFECTION

Cats have a highly developed sense of touch. Their skin is covered with sensitive nerves and responds to the slightest pressure. When in the right mood, the cat enjoys being petted.

A cat shows affection by rubbing against the legs of a friend.

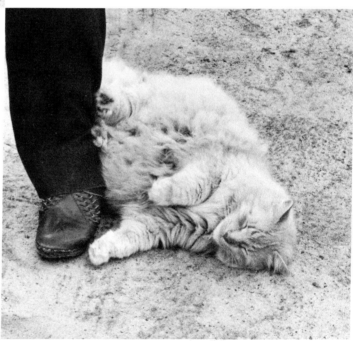

Sometimes when it wants attention, it turns over, asking to be scratched.

A boy gets an unexpected kiss from his cat.
The cat's kiss is a lick or a touch with its nose.

SMELL

An acute sense of smell is important to a cat's life. It is used to find food, hunt, identify friends, enemies, and whatever else a cat may encounter.

The cat enjoys a walk through grass and weeds. There is a lot to smell and nibble in a backyard jungle.

Especially nice is the smell of this flower. Many different plants and flowers attract cats.

The catnip plant is a favorite. Cats sniff, eat, and roll in it with obvious pleasure.

HEARING

The range of a cat's hearing is enormous—much wider than a dog's or a human's. Many cat owners tell "incredible" tales of how their cats know they are coming home long before they enter the street or approach the house. Often a cat will leap to its feet when, in seemingly complete silence, some sound apparently reaches its ears—a sound impossible to describe because no human can hear it.

With a faraway look in its eyes, the cat listens to a distant noise. Its highly movable ears have ridges that pick up the sound and establish where it comes from.

The Body

The body of a cat is extremely flexible. Most unusual is the cat's walk. The cat can move front and hind legs on the same side of the body at the same time. No other animal can walk like that except the camel and giraffe. This movement helps the cat to walk on extremely narrow surfaces.

The flexibility of the spinal column makes it possible for the cat to arch its back, elongate or contract it.

With tongue or moistened
paws, the cat twists
to wash every part of
its body except between
the shoulder blades.

Not so much as an owl
but more than other animals,
the cat can turn its head
to look behind itself.

A cat rests comfortably on a glass table with its sharp claws out of sight.

PAWS AND CLAWS

Forepaws and claws are very important parts of the anatomy. A cat uses the forepaws to clean its fur, climb trees, touch and identify objects, pull out things from inaccessible places, play or fight, and defend itself. The hind paws are not as flexible but strong and well used when jumping or running. Most of the time, claws are retracted to keep them sharp and protected from injury.

With the help of elastic ligaments, the claws can be pulled back or extended. Cats often exercise paws to keep those ligaments flexible.

A cat tiptoes silently on soft pads with retracted claws.

The cat uses the tree as a scratching post to remove worn-out protective sheathing from the claws of the forepaws. It chews off the worn sheathing from the hind paws.

Escape

The cat is not a long-distance runner. In short sprints it can outrun a pursuing dog—but not for long. The cat soon gets tired and has to look for safety high up or deep down.

Its hind legs are in unison when the cat runs at full speed.

Reaching a hiding place the cat "elongates" its back, "shortens" its legs, and goes underground.

Climbing a tree is the surest way to escape a pursuer.

When there is no time
to go up or down,
a garden rack is better
than nothing.

Whiskers

Whiskers and eyebrows are connected to highly sensitive nerves and rich blood supply and so are additional organs of touch.
In poor light, even in total darkness, they play an important part in cat orientation.

When entering a dark confined place, a cat uses its whiskers to help judge the surroundings.

The Tail

A cat's tail has a language of its own. Depending on the cat's mood, it lashes, twitches, flips, weaves, jerks, rises high, or sinks low. A cat owner understands the talk of the tail.

Extremely flexible, the tail is arched high over the cat's back during play.

When the cat is resting, the tail seems to relax.

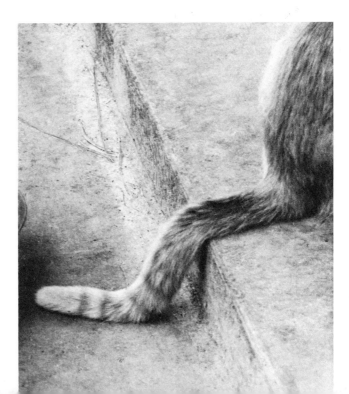

Voice

Using a wide variety of sounds, a cat can express affection, anger, likes, dislikes, complaints, and delight. A soft purr and a loud call that can be heard over a long distance are only parts of the extensive sound vocabulary a cat uses to tell about itself.

A tender meow is reserved for friends—human or feline.

There is no limit to the hisses, screams, and snarls in an angry cat's language.

A loud demand leaves no doubt that the cat family is hungry. Cats like their meals served on time.

With a deep growl and ears flat back, the cat makes known its displeasure.

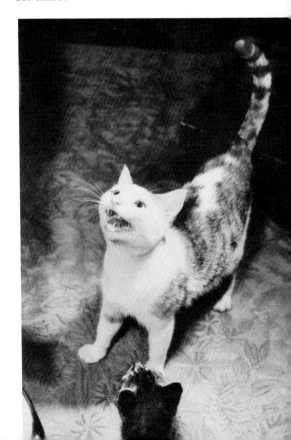

Cat Mothers

Cats select their "maternity wards" carefully. A dark and dry place with a soft floor is choice. A place where the cat has slept is often chosen by the future mother.

This cat, pet of a Latin music band, nurses the kittens between the band instruments—her favorite corner.

A mother cat living on a farm brings up her second litter of kittens in a hayloft, under the roof of the barn.

A mother cat having a single kitten often nurses the kitten much longer than a mother with a litter of two or more kittens.

The bond between the mother and its one kitten can last for many months. Most cats are devoted and patient mothers, regardless of how many kittens there are.

Waiting for Mother's Return

The first weeks after giving birth, the mother cat stays with her kittens almost continuously; only short trips to get food or use the litterbox interrupt the nursing. After the fifth week, the kittens are left alone for longer periods of time. They are starting to eat food other than mother's milk, and will soon be on their own.

Kitten at Play

There is nothing more amusing
and graceful than a kitten at play.

A kitten plays with anything
that swings, rolls, or jumps—
the yo-yo is perfect.
Playing with utmost energy and
concentration, it soon gets
tired and has to take a nap.

The kitten will often use its sensitive front paws to investigate a "mysterious" object.

Intelligence

Scientists have proved that cats have a high level of intelligence. Even young kittens can solve problems that require quick thinking.

A six-week-old kitten runs in a maze during an experiment to measure its learning ability. In no time at all, the kitten figured out the route to the food box.

Curiosity is a part of a kitten's nature and is never outgrown.

Always expecting excitement,
a kitten never misses
an opportunity to play.

They chase, hug, and pounce
on each other . . .

. . . or follow a ball of knitting yarn, to catch it. As long as the kitten plays and does not nibble it, the yarn is harmless. Swallowed, yarn can be very harmful.

An open door can mean fun or trouble. Either way, the kittens are tempted to find out.

Courage is all that is needed to make the jump.

After climbing up the back of a chair, the kitten sees it must go down.
It is not an easy decision to make. The kitten takes time to think it over.

Landing on Its Feet...

Contrary to popular belief, a falling cat cannot always right itself and land on its feet. Cats do get killed when falling from high places. Even from lesser heights, a fall can result in severe injury. For a sleepy cat, a fall from any window could be fatal. Cats usually judge where and when it is safe to jump. Sometimes, if they find they have ventured too high, they wait to be rescued.

A perfect dive, and the kitten lands on its feet.

A little shaky for a moment . . . but it was worth the trip.

In a Mirror a Kitten Sees a Stranger

A kitten looking in a mirror sees another kitten.

In a distortion mirror, the strange kitten looks very unusual.

The kitten is puzzled—
it is not sure what kind
of a playmate this is.

The best way to find out
is to get close and try
a kiss—nose to nose.

Way of Life

Adult cats spend most of their time resting and thinking. They never hurry if not pursued, take advantage of comfort where they can find it, and appreciate it when people help them to live their own way.
This tomcat stays home till nightfall, then disappears into the woods.

Hours are spent meditating on top of discarded newspapers in the basement of an apartment house.

The roof is an ideal observation post. From high places, cats can observe movements in high trees, on low ground, and the curious activities of neighborhood cats.

Investigating an old, deserted barn is a challenge no cat can resist.

Hunting Habits

Even pet cats that don't have to hunt for food don't lose the ancient hunting habits imprinted on them —to watch, pounce, and catch a prey.

With infinite patience, this cat watches the birdhouse.

Resting under a car, the tabby is alert and waiting.

Camouflaged under a tree, a tortoiseshell sits motionless, hoping some unsuspecting prey may appear.

Cats Love Warmth

The cat always finds a place
in the sun. It will settle
down on a log or, after
some stretching and turning,
fall asleep on a sun-heated rock.

Inside the house,
a favorite place is
under a desk lamp.

The young Siamese found
a warm spot on the rug
where the sun shines
through the window.

A Cat Is Never Too Old to Play

The adult cat plays very much the same way as a kitten does—everywhere and with everything. For a house cat, play is the only exercise it gets. It plays often and with more energy and enthusiasm than does an outdoor cat. The cat invents games and is always full of new ideas.

A pencil is a newly discovered toy.

A shopping bag is a good place to hide, then to jump out and catch something that only a cat can see.

This tabby is a champion string player.
The white cat watches, but does not participate.

Bored with the show,
the white cat decides to leave.

Smart Learners

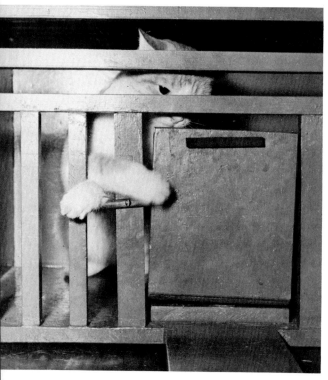

In this experiment, a cat
is faced with a problem
and solves it. All cats
have high levels of intelligence—
only the degree of smartness
and their personalities differ.

With crossed paws, a smart cat
establishes leverage and gets
the door open.

When the door drops down,
the cat rushes out
to get the prize—
its favorite food.

This Siamese cat remains motionless and will wait till the cup and saucer are removed.

Some cats learn to do tricks not because they have to but because they want to. Some tricks they learn entirely on their own. They may not be spectacular, but the owners are delighted.

This cat sits up expecting a treat.

Swimmers

All cats can swim—some like to and some don't. Some enter the water to fish or when chased by dogs, others do it for the pure pleasure of swimming.

This cat takes a boat ride.
To relieve the monotony, he leaves the boat for a swim.

Like a torpedo, he cuts through the water.

The empty bathtub is a favorite resting place.

A SWIM IN THE BATHTUB

As a kitten, this cat accidentally fell into a bathtub filled with warm water. He liked it so much that every time the bathtub is filled, he tries to jump in. He is not allowed to do it—but occasionally, the bathtub is filled especially for him.

A swim around the bathtub is enjoyable.

Against his will, he is lifted out and given a rubdown with a towel so that he won't catch cold.

Sleep

Cats are nocturnal animals; they should sleep all day and walk all night, but living as house cats has changed the sleeping habits of many of them. They move around all day and sleep all night. Cats like to sleep. Whenever there is nothing better to do, they close their eyes and go to sleep—or just pretend to.

Despite the usual demands for softness and comfort, the unpredictable cat takes a nap in a hard metal wheelbarrow.

This cat selects a soft place, where it sleeps in a most unusual position.

Some cats have preferences: a certain chair, a certain pillow, a certain toy. This cat will not sleep without her doll.

Cats and Friends

Cats like the company of other animals. They form lasting friendships with big and small mammals. A tame rat, an elephant, ducks, parrots, and pigeons are among cats' friends.

The friendship of cats and horses is legendary. In many stables the horse has a favorite cat and the feeling is always mutual.
They often sleep close by and the cat follows the horse to the meadow.

A cat and a dog who share a home or grow up together are devoted and playful companions. Cats and dogs are not "natural enemies"—they can be friends or enemies depending upon individual personalities and circumstances. A cat may attack a strange dog in defense or to protect her kittens. A past unpleasant experience can also make a cat distrustful of dogs.

Adoption

A mother cat is always willing to nurse a young animal— even if it's a squirrel, rabbit, wolf, or a dog.

When a motherless puppy was added to her five kittens, the cat mother adopted it. The puppy was larger than the kittens, but that did not seem to disturb anyone in the cat box.

The cat accepts the grown-up dog she has nursed as a different-looking, legitimate member of her family.

The farm cat meets all kinds of animals. To be nuzzled by a calf may be a thrilling experience for the house cat but the cat on a farm thinks nothing of it.

Purebred Cats

Pedigree breeds are the aristocracy of the cat world. They are exhibited, given awards, and praised highly. Many breeds are named after countries in which they supposedly originate. In most cases the breeds cannot be found in those countries. They are usually bred in Europe or the United States. But even the rigorous beauty standards cannot change the cat's nature. Purebred or common, it's the same independent *Felis catus*.

The Abyssinian, with its slender, graceful body, resembles cats of ancient Egyptian art.

The white Persian, a long-haired cat, has for many years been favored by cat lovers everywhere.

The Siamese, the most popular of all purebred cats, likes human company, is vocal, and appears exotic and mysterious. It is the only domestic cat with nonretractable claws.

A Siamese mother nursing her kittens. When born, they are almost white.

As the kitten grows older, the fur on its face, ears, legs, and tail becomes darker.

The Siamese became a symbol of luxury.
According to one story, they were first bred by a king of Siam
and lived in the royal palace, then were imported by English breeders
about a hundred years ago, and became favorites of European society.

The Himalayan is a new breed. It took a long time to breed a perfect Himalayan—a cross between a Siamese and a Persian—but the result was worth waiting for.

The Manx, a cat without a tail, has features in common with the rabbit: very high hind legs and typical rabbit gait, a double coat, and a short thick undercoat. Despite the resemblance, the Manx is not rabbitlike. It's an expert climber, a courageous hunter and fisher—and an intelligent pet.

The Burmese originated in France, where the breed was called Sibeline. The slim, elegant cat with golden eyes and a shiny, solid brown coat is smart and inquisitive.

The calico is red, black, and white. Almost all calicos are females. Sailors are usually fond of cats and keep them on ships as mascots. Japanese seamen believe the calico brings luck, keeps away evil spirits, and prevents shipwrecks.

The Maine coon cat is a sturdy, big cat with semilong hair, able to withstand very cold weather. As for its origin, one source says Queen Marie Antoinette sent some of her Persian cats to a refuge in Maine during the French Revolution. There they mated with local cats—and so was the Maine coon cat born. More probably sailors from Maine, in the last century, brought long-haired cats home from the Orient and, after mating them with native tabbies, produced the first Maine coon cats. There is no relation between the coon cats and the raccoon.

Two Different Lives

The house cat, whose days are spent playing, resting, and being admired, has an easier life.

The tomcat, a tough character, is always looking for challenges.
Scars and torn ears tell of nightly adventures.

ABOUT THE AUTHOR

NINA LEEN, former *Life* photographer, has written extensively about animal life. Among her celebrated books are *And Then There Were None, The Bat,* an ALA Notable Book for 1976, *Snakes,* an ALA Notable Book for 1978 and an Outstanding Children's Science Trade Book chosen by the National Science Teachers Association, and *Monkeys.* Ms. Leen lives in New York City.